Marcos B. S. Martins
Liliany Viana

Study of the Bioecology of the Mapará (Hypophthalmus marginatus)

Marcos B. S. Martins
Liliany Viana

Study of the Bioecology of the Mapará (Hypophthalmus marginatus)

in the municipality of Cametá-PA, Cametá micro-region

ScienciaScripts

Imprint
Any brand names and product names mentioned in this book are subject to trademark, brand or patent protection and are trademarks or registered trademarks of their respective holders. The use of brand names, product names, common names, trade names, product descriptions etc. even without a particular marking in this work is in no way to be construed to mean that such names may be regarded as unrestricted in respect of trademark and brand protection legislation and could thus be used by anyone.

Cover image: www.ingimage.com

This book is a translation from the original published under ISBN 978-613-9-62662-5.

Publisher:
Sciencia Scripts
is a trademark of
Dodo Books Indian Ocean Ltd. and OmniScriptum S.R.L publishing group

120 High Road, East Finchley, London, N2 9ED, United Kingdom
Str. Armeneasca 28/1, office 1, Chisinau MD-2012, Republic of Moldova, Europe
Printed at: see last page
ISBN: 978-620-7-74803-7

SUMMARY

To my parents, **Olenilson and Maria Laide**, for their example and tireless encouragement.

My grandparents, uncles and aunts, cousins, who have always strengthened my journey with noble words and gestures.

And to everyone who has contributed in some way to this victory in my life.

Liliany Lobato Viana

To my parents, **Ray Martins and Edilma Martins**, for the opportunity they gave me. To my siblings Cris Martins, Cristiane Martins and Kérina Martins and my sister-in-law Cinthia Martins, who have always focused on my success. And to everyone who has contributed in some way to this victory in my life.

Marcos Benigno Silva Martins

First of all to GOD for his omnipresence at every moment of my life.

To my parents, Olenilson and Maria Laide, who always encouraged me and fought for me to have the best education possible, whose love above all other goals is unconditional, and who strengthened me in every battle, whatever the outcome, thus building the basis of the human being I am today.

To my sister, Lanaide, for her love, friendship and unconditional support.

To my friend and advisor Professor M.Sc. Gyanne do Socorro Pereira de Lima, who guided my steps in the decisive phase of my academic life and helped me to fulfill all the objectives set for this work.

To my friend and fellow TCC student, Marcos Martins, for all the difficulties we have overcome together.

To my class friends, especially: Andren, Camila, Hadriane, Janaina, Junior, Kâtia and Tatiane, whose demonstration of companionship has been countless during these years, invigorating me in difficult times and cheering me up on happy occasions that will never be forgotten.

To my siblings, friends and companions who have lived with me in the republica, especially: Adila, Diane, Fabielle, Eliane and Luana, who in the last year have shared difficult moments that we have overcome together and many other moments of happiness and true friendship, which will be forever remembered.

To my friends Assis, Gilson, Joelson, Leo and Roberto, whom I met over the years in Cametà, and who never lacked friendship, solidarity and support.

To my friends in Abaetetuba, who, even though far away, never lacked friendship, solidarity and support.

To the fishermen Mr. Pantoja and Mr. Joelite Maria, from the Jaracuera community, for their help during data collection.

To the President of the Z-16 Fishermen's Colony, Mr. José Fernando, and to the employee Dinho, for their collaboration with this research work.

To Mr. José Danùzio, from IBGE, for the theoretical support he provided when it was most needed.

Liliany Lobato Viana

First of all, I thank GOD for having ratified my conduct at this milestone in my life.

To my advisor, Professor M.Sc. Gyanne Lima, who guided my steps in the decisive phase of my academic life and helped me to fulfill all the objectives set for this work.

To my esteemed work partner and friend, who knew how to understand the moments when difficulties were at my side.

To my friend Anderson Baia for his gracious didactic contributions to the success of this work.

Andren, Camila, Eliéte, Deylane, Edvan, Glàucio, Hadriane, Kâtia, José Jr, Janaina, Marcelle, Marquinho, Paulo, Renato and Tatiane.

To my friends and housemates Serginho and Rodrigo, who were crucial to my organization.

To my friends Leo, Gilson, Prof°. José Maria, Frank, and so many others that I cultivated during the years I spent in the city of Cametà studying.

To the fishermen Mr. Pantoja and Mr. Joelite Maria, from the Jaracuera community, for their help during data collection.

To the President of the Z-16 Fishermen's Colony, Mr. José Fernando, and to the employee Dinho, for their collaboration with this research work.

To Mr. José Danùzio, from IBGE, for the theoretical support he provided when it was most needed.

To everyone who contributed directly and/or indirectly to this achievement.

Marcos Benigno Silva Martins

"The wisdom of a human being is not in how much he knows, but in how much he is aware that he doesn't know."

(Augusto Cury)

VIANA, Liliany Lobato & MARTINS, Marcos Benigno Silva. **Study of the bioecology of the maparà *(Hypophthalmus marginatus* Valencienes 1840, Siluriforme) in the municipality of Cametà - PA, Micro-region of Cametâ.** 49p. Graduation (Full Degree in Natural Sciences - Biology) Para State University, 2008.

SUMMARY

The Tocantins River is part of the Amazon Estuary. With the closure of the Tucurui HPP in 1984, the river suffered major flooding. The mapara is a migratory species of fish from the catfish family. They look for suitable environments to reproduce, spawn and/or feed, and generally live in shoals. They are medium-sized, especially those found in the lower and middle Tocantins River. It is believed that the dam has influcnood the behavior and volume of the shoals of the mapara *(H. marginatus).* The aim of this study was to analyze the situation of mapara from Cametâ, in the Cameta micro-region, assessing the species' bioecology in the region. The methodology consisted of bibliographical research into the biology and ecology of the species, and interviews with fishermen in the region, mainly in the Jaracuera community. In order to investigate the current state of the mapara catch in the region, morphometric data (FC) was obtained from fish markets and fairs in the municipality. The results confirmed that the juveniles (12 to 24 cm furcal length - FC) caught are below the size required for reproduction (37 cm), and are often caught during the species' closed season (November to February). There is a difference in the body color pattern of *H. marginatus* from the Cametâ micro-region compared to those from other regions. The mapara (*H. marginatus)* found at the mouth of the Para River up to near the Tucurui dam is popularly known as "Mapara-do-nosso", they live in lentic environments, are smaller, robust and the color of their backs is a lighter gray, while the mapara found both along the course of the Tocantins River above the dam and in the dam lake (lentic environments) are elongated and narrow with dark backs, and are popularly known as "Paulo Isidòrio". It can be concluded that the environmental changes that have taken place in the Tocantins River with the construction of the Tucurui Hydroelectric Power Station, plus the improper catches during the closed season and of juvenile individuals of the species *H. marginatus, the fishing of mapara and the fishing of* the species *H. marginatus. marginatus,* the mapara fishery could enter a state of overfishing, which could also cause serious bio-ecological and socio-economic changes, since the Cametaense population depends fundamentally on fishing in the Tocantins River, especially the mapara, which is traditional for the region.

Keywords: Mapara, bio-ecology, Tocantins River, Cameta - PA.

VIANA, Liliany Lobato & MARTINS, Marcos Benigno Silva. **Study of the bioecology of the maparà *(Hypophthalmus marginatus* Valencienes 1840, Siluriforme) in the municipality of Cametà - PA, Micro-region of Cametâ.** 49p. Graduation (Full Degree in Natural Sciences - Biology) Para State University, 2008.

1. INTRODUCTION

The great diversity of ecological environments in South America has allowed for a great deal of evolutionary radiation. Today, this region has a very rich fish fauna, with around 60 families and approximately 2,800 known species. It is estimated that there are around 8,000 species of Neotropical freshwater fish. This extremely complex fauna is, from an evolutionary point of view, a great product of the biological world. Fish species often subdivide into isolated local populations. This phenomenon is more pronounced in freshwater fish that inhabit rivers or lakes with physical barriers, while marine fish populations are less delimited. Therefore, historical factors such as colonization and the emergence of physical or geographical barriers can influence the distribution of fish (HOWEL & BLACK, 1980).

The species richness of the Amazon basin is well known. Bohlke *et a.* (1978) estimated that the total number of species present in the basin is around 2000, 3% of which is still unknown. According to Roberts (1972) 1,300 species have already been described. Of the species described, 300 have been recorded in the Tocantins River alone, in just 500 km of its lower course (SANTOS *et a.*, 1985).

The Amazon basin is made up of several rivers that are home to large numbers of people, whose livelihoods and economies depend mainly on the resources produced by these rivers. Fishing is an economic activity that generates employment and income, and is traditionally important for these populations, providing the main source of animal protein for the poorest sections of society (PETRERE JR. 1992, *aput* ALCÂNTARA-NETO, 1994). Agostinho (1992) points out that damming a river leads to instability in the downstream stretch, a reduction in the diversity of the local ichthyofauna in the flooded stretch and the confinement of species in the upstream stretch.

Fishing in the Amazon has an artisanal character and even so, fish is the main source of protein for human consumption, particularly for the people who live on the banks of the region's rivers and lakes (SANTOS, 2006).

Among the rivers of the Amazon estuary, the Tocantins River is of paramount importance, even though it has been subjected to a strong environmental impact due to the closure of the Tucurui HPP in 1984; flooding an area of 2,430 km^{2}, according to calculations based on satellite images (FEARNSIDE, 1999). The HPP was built as a result of the growing demand for electricity in the country, driven by bauxite mining. Despite the environmental, social and economic impacts caused by the damming of the Tocantins River, an important alternative for subsistence and income generation for the population bordering the reservoir remained artisanal fishing. However, fishing activity was drastically reduced downstream of the dam, in the lower Tocantins region.

Fishing in the area of influence of the Tucurui Hydroelectric Power Station (UHE Tucurui) is an activity of great economic and social importance, especially for the municipalities located upstream and downstream of the dam: Marabà, Nova Ipixuna, Itupiranga, Jacundà, Breu Branco, Tucurui and Baiâo, upstream, and Mocajuba, Cametà, Limoeiro do Ajuru, Igarapé-Miri and Abaetetuba, downstream. Many jobs are generated by this activity and it is estimated that around 10,000 fishermen work downstream and upstream of the dam, which represents a human population of approximately 50,000 people who depend directly or indirectly on fishing (JURAS; CINTRA; LUDOVINO, 2004). According to Clark (1990), the density of fishermen working in the reservoir, upstream and downstream, is increasing by 10% per year.

The Amazon is a traditional fishing region, ancient and highly developed, at least in certain areas. In the 1970s, questions arose among fishing communities and the general public about the possibility of overfishing. According to Mérona (1995), this questioning was based on empirical observations related, for example, to the decrease in the average size of the catch of some species that were prized on the markets or the increase in the distance traveled by boats to bring in these species.

As with any resource exploitation activity, fishing takes into account economic, technological, cultural and political aspects that add to the complexity of "fishing" systems. However, the biological and ecological aspects are fundamental to management, as obviously there can be no exploitation without resources (MÉRONA, 1995).

According to Mérona (1990), fishing significantly modifies the composition and probably also the structure of aquatic communities. When the dam was closed, Wootton (1984. p. 1-12) said that: "The success of a reproductive strategy guarantees the survival of the greatest possible number of offspring and represents the product of a long process of natural selection and adaptation to seasonal oscillations in environmental variables."

The closure of the hydroelectric dam in September 1984 had immediate consequences for fishing (MÉRONA, 1990). Firstly, the total interruption of the river's flow during the filling of the dam caused the water to become stagnant and the back and forth movement of the tides favored the incorporation of a large mass of sediment which, in turn, caused a phenomenon of "self-eutrophication". The mobile animals would have escaped from the area. At the same time, the fishermen maintained a high level of fishing pressure, bringing large quantities of juvenile maparà *(Hmarginatisi)* to the main markets in the Cametâ micro-region, caught at the time of the upward migration.

It is clear that the Tucurui HPP has directly and indirectly had a significant impact on commercial fishing. However, it is hilarious to note that they are also the fruits of the consequences of deforestation (CUMMINIUS *et al,* 1985; WORD & STANFORD, 1989). Marginal vegetation influences light and temperature cycles, controls bank erosion and enriches the aquatic environment by introducing allochthonous organic matter.

The maparà (*H marginatus)* continues to be the most important species for commercial fishing in the Tocantins River, in the area of influence of the Tucurui HPP. The drop in production in the downstream region seems to have been offset by the increase in production in the reservoir area, which occurred after the limnological conditions in that stretch stabilized. It is possible that the high availability of nutrients from the decomposition of organic matter in the reservoir area, as well as the greater incidence and penetration of light in that area, resulted in greater local primary productivity. As the maparà is a planktivorous species, it can take direct and rapid advantage of the increase in food supply with the formation of the lake, resulting in the proliferation of the species in that environment. Efrem & Jansen (2000).

With the advent of the hydroelectric dam, the situation of fishing culture has become critical in terms of commercial productivity in the areas downstream from the Tucurui HPP. This is confirmed by the poet Raimundo Castilho e Silva (1986), a fisherman from the Tocantins region, in his poem **"A pesca do Maparà"** :

> "[...]
>
> XXIII
>
> Today's fishermen
>
> No calm, no ideal.
>
> Being very ambitious, they lead themselves astray Thinking that Maparâ is a mineral product XXIV
>
> They must know that this fish
>
> You must have salt freedom
>
> You have to spawn
>
> With great peace of mind If not, he'll get rid of it with the greatest of ease.
>
> XXV
>
> The stubborn fishermen
>
> They don't want to know
>
> Driven by greed

They renounce their duty

They catch fish at any time They're ready to sell. [...]

XXVIII

He never arrived

Life-size

You can see that the boatmen

On any and all coastlines They don't allow it to grow And that way it would be cooler. [...] "

Habitat modification directly influences the biological functions of each fish species (feeding, reproduction, growth and mortality) and ecology (migration). However, species are not isolated; they are part of complex biocenoses and changes to any one element also affect the interrelationships between species. The end result of this set of interconnected changes is evidenced by the characteristics presented by the ichthyic communities, Mérona (1985). In this context, the Tucurui HPP is important because it was the first major dam to be built in the Amazon, and because it severely affected existing fish habitats, such as the mapara. Therefore, the purpose of this study is to analyze the migratory pattern of the species *H. marginatus* downstream of the dam, making comparisons with migratory fluctuations upstream and even in the area delimited by the reservoir. And a detailed discussion of post-filling fisheries production. There has been a drastic reduction in commercial fishing catches in the Cameta region. The evolution of commercial fishing catches in this region between 1981 and 1998 shows that there has been a steady decline in fishing production, which fell from 900 t/year in 1981 to 492 t/year in 1998, a drop of 89%. It is possible that this drop in catches is the result of changes in the river's hydrological regime after the dam was built, since the rhythm of flooding and ebbing was now regulated by the HPP's operating activities. This may have disoriented and modified the migratory behavior of the shoals of some of the main commercial fish species in the region, contributing to the decrease in catches (EFREM & JANSEN, 2000).

In the Cametà region, the maparà fish has undergone changes in reproductive habits. After the reservoir was filled (1985-1986) (INPA, 1986; COLLART, 1986), there was a decrease in fish production downstream of the dam. This can be explained by various factors, such as: (1) changes in the flood regime, which reflects on the environmental conditions for fish reproduction; (2) impoverishment of the flowing water from the retention of nutrients by the reservoir; (3) due to the plant's operating rules, especially during the dry season, the water that passes into the downstream stretch comes from the bottom of the reservoir (anoxic layer), where there is an increase in conductivity, total suspended solids, iron, and especially phosphate nutrients and ammonia (CET, 1988); (4) due to the alterations to the hydrological cycle caused by the damming of the river and the operation of the Tucurui HPP, there is a high mortality rate of eggs, larvae and fry, which compromises recruitment and the replenishment of fish stocks. The decline in stocks is undesirable from an economic, ecological and social point of view, and in order to avoid it, it is necessary to produce scientific knowledge that allows the situation of fishing resources to be diagnosed and makes it possible to adopt measures for the efficient management of these resources, This knowledge covers the most diverse areas in which all the processes linked to this activity take place, such as socioeconomics, bioecology, environmental education and fishing technology, among others.

From this perspective, this research project was conceived, aimed at studying the bioecology of the maparà *(H. marginatuS)* 24 years after the closure of the Tucurui HPP, taking into account the knowledge and experiences of fishermen with more than 25 years of work.

2. GENERAL CONSIDERATIONS ABOUT THE MAPARA

2.1 TAXONOMY

Depending on where they occur, maparâs *(Hypophthalmus ssp)* can be found under the vernacular names: mapurà, peixe-gato, olho-de- gato, lau-lau, mapa-racuii, moça, mandubé rosado, luz baixa, salàrio and mapará de cametâ (PEDREIRA, 1971; IBGE, 1989). However, the most peculiar name is mapará. They are bony fish, belonging to the class Teleostomi; of the Super Order Ostariophysi; of the Order Siluriforme (ALCÂNTARA-NETO, 1994). Until recently, they were part of a separate family called Hypophthalmidae, which has now been included in the Pimelodidae family (SANTOS, 2006). They have the following taxonomic classification:

Superclass - Gnothostomata

Class - teleostomi

Subclass - Ostheichthyes

Infra Class - Actinopterygii

Division - Teleostei

Superorder - Ostariophysi

Order - Siluriformes

Family - Pimelodidae

Genus - *Hypophthalmus*

Species - *Hypophthalmus edentatus* Spix (1829)

Hypophthalmus marginatus Val. (1840)

Hypophthalmus fmbriatus Kner (1857)

2.1.1 The siluriform order

This order is characterized by being made up of fish without scales on their bodies, covered by bare skin or bony plates, commonly known as leather fish or cascudos, with barbels around their mouths usually in three pairs, one pair being maxillary and two pairs being mentonian. They have small, curved teeth grouped together in plates or bands with a sandpaper-like appearance. The pectoral and dorsal fins are usually garnished with spines with saws on the margins (BRITISKI *et al.,* 1984: FERREIRA *et al,* 1998).

The Siluriformes are one of the most diverse and widely distributed groups of Neotropical fish among the Ostariophysi, comprising a total of 34 families and 412 genera, distributed over approximately 2,405 species (NELSON, 1994). In the freshwaters of South America, the Siluriformes comprise 14 families, 12 of which inhabit the Brazilian Amazon (FERREIRA *et al.,* 1998), making up a total of 40% of the species classified for the group. According to Rapp Py - Daniel & Fernandes, (2005) among the Siluriformes families, only the Plotosidae family is exclusively marine, and the Ariidae family has both marine and freshwater representatives, and the vast majority of these families occur exclusively in freshwater.

2.1.2 The family hypophthalmidae cope 1871

According to Oliveira (1981) and Santos *et al.* (1984), this family, with only one genus and only three species, is very peculiar within the catfish groups, because while most catfish are dermal and carnivorous, the Maparâs are closely pelagic and planktophagous. Mainly due to this last adaptation, it has a very wide gill opening and terminal mouth; long and numerous gill tracks; the eyes are relatively small and are located on the lateral part of the skull, close to the mouth commissures. Maxillary and palatine teeth absent, pharyngeal teeth present. Lateral line with numerous upper and lower branches; atrophied and encapsulated swim bladder. The rays of the fins are all soft; the anal fin is long and the tail is pierced or truncated; ventral and adipose fins are reduced. Three pairs of well-developed barbels, one pair of maxillary barbels and two mentonian barbels; the maxillary barbels may have very wide posterior fimbriae. The head is long and flattened, depressed anteriorly and high posteriorly, with a greatly expanded ethmoid bone; the bones of the head are covered in fur. The body is elongated, laterally compressed, devoid of scales or plaques and covered in fine skin. The general coloration of the body is metallic lead grey, with the body darker than the belly.

Until recently, maparàs *(Hypophthalmus spp)* were grouped into the Hypophtalmidae family. However, with the changes that have taken place over the years, maparàs have been structured into a new family called Pimelodidae, which currently includes 83 species, distributed among 80 genera. However, it is estimated that there are around 45 undescribed species (REIS *et al* 2003). Its representatives are exclusively freshwater and are distributed throughout South America and Panama (LUNDBERG & LITTMANN, 2003).

2.1.3 The genus *hypophthalmus* spix

The fish of the Pimelodidae family, which are relatively common in the Amazon region, are grouped into a few genera. According to Oliveira (1981), the genus *Hypophthalmus* belongs to this region. This genus can be described based on the following characteristics: Dorsal fin situated just before the middle of the body (excluding the tail), at the level of the anterior part of the anal fin, which is very long; very small ventral and adipose fins, the first situated anterior to the dorsal fin (anus also anterior to the dorsal fin) and the second on the last rays of the anal fin; pectoral fin reaching, sometimes surpassing, the origin of the anal fin; caudal fin pierced with tapered lobes, or notched (or emarginate) with truncated lower lobe. Acuoleus of dorsal and pectoral fins serrated and rigid in juveniles, smooth and flexible in adults. Straight lateral line with numerous parallel oblique upper and lower branches directed towards the rear; the appearance of the branches on the lateral line is generally not uniform, and gaps are common, especially in juveniles; two rows of pores flank the lateral line and its branches. Posterior end of the lateral line curved upwards over the upper caudal lobe. The lateral line and its branches are internally lined with a rigid, helical fibrous tissue. As it develops, bone is deposited on the lateral line, forming a row of short tubules. The barbels do not follow the growth of the animal, but are proportionally larger in younger animals.

2.1.4 The *hypophthalmus* species

The genus *Hypophthalmus* is characterized by only 3 (three) species: *H. edentatus* Spix (1829); *Hmarginatus* Val. (1840), found in the lower Tocantins region; and *H fmbriatus* Kner. (1857), which has the following description:

A) *H. edentatus* Spix (1829)

According to the redescription by Oliveira (1981), it has a slightly notched caudal; tapered upper lobe and rounded lower lobe, sometimes truncated, caudal edge and often dark anal edge; pectoral ends always extend beyond the origin of the anal; relatively short head. Mentonian barbill with relatively narrow fimbria, with its greatest width close to the base, never reaching the end of the pectoralis; the width of the fimbria rarely exceeds ½ diameter of the orbit; maxillary and mentonian barbills approximate in size and shape; the buccal commissure that continues after the corner of the mouth never reaches the anterior level of the orbit; lateral line with 25 or fewer branches (or corresponding spaces) between the level of the end of the dorsal and the level of the end of the base of the adipose, see Figure 1.

Figura 1: Species of Maparà, *H edentatus* Spix (1829)

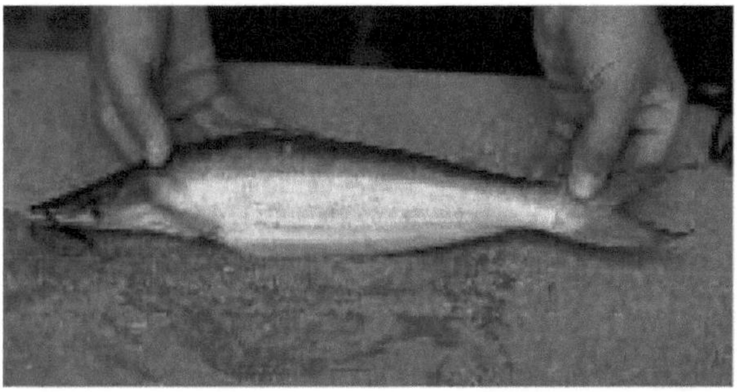

Source: Google image.

8) *H marginatus Val* (1840)

According to the redescription of Carvalho (1978a); Oliveira (1981); Santos *et al* (1984), it has a spindle-shaped, elongated body, metallic bluish-gray on the back and bright yellowish-white ventrally. Head flattened dorso-ventrally, fontanel wide, longitudinal; anterior terminal mouth wide; premaxillae and dentary teeth without teeth, showing aciculae on mandibular plates. Small eyes in an almost ventral position; double lateral nostrils, six barbels extending to the insertion of the pectoral fins. Caudal fin homocercal, deeply furcated, lobes tapered, in general the upper one slightly longer and always narrower than the lower one; edge of caudal and often anal fin dark; pectoral fin located on the anterior sixth of the body, next to the opercular opening, its tip rarely reaching the origin of the anal fin; rudimentary ventral fins, located anteriorly, inserted a few centimeters from the anal opening; dorsal fin located in the anterior third, quite far from the occipital process; long, multi-grooved anal fin, with its base located just after the anus. Head relatively long; maxillary and mentonian barbels similar in shape, with the maxillary usually slightly larger than the mentonian; fimbria of the mentonian barbrel relatively narrow, with its greatest width near the base, rarely exceeding ½ diameter of the orbit; the end of the mentonian barbule does not reach the end of the pectoralis in specimens over 60 mm in standard length; the mouth commissure that continues after the corner of the mouth reaches the middle level of the orbit inferiorly; lateral line with 25-29 branches (or corresponding spaces) between the level of the end of the dorsal base and the level of the end of the adipose base.

Figura 2: Mapará species, *H marginatus Val* (1840)

Source: Viana, L.L.

C) *H. fmbriatus* Kner. (1857)

The caudal fin is deeply furcated; tapered lobes, in general the upper one slightly longer and always narrower than the lower one; hyaline caudal and anal edges; pectoral ends always reach the origin of the anal, relatively short head. Mentonian barbel much larger than the others, always exceeds the end of the pectoralis; fimbria of the mentonian barbel much wider than the others, with its greatest width in the distal portion, always exceeds the diameter of the orbit (usually double); mentonian barbel strongly pigmented (black) from the juvenile form; the mouth commissure that continues after the corner of the mouth reaches, inferiorly, the middle level of the orbit; lateral line with 25-26 branches (or corresponding spaces) between the level of the end of the dorsal base and the level of the end of the adipose base. Alcântara-Neto (1994)

Figura 3: Species of Mapará *H fmbriatus* Kner. (1857)

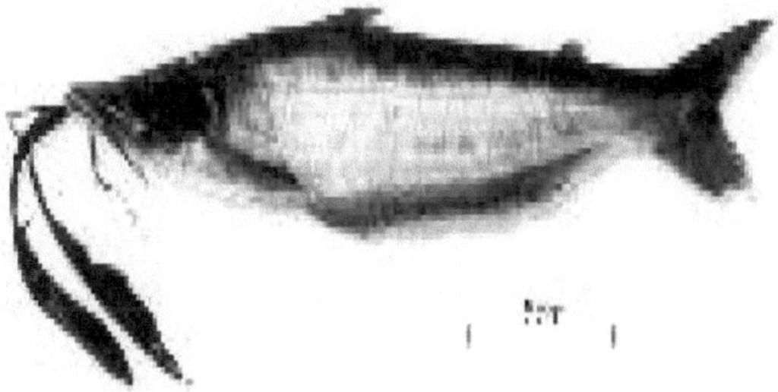

Source: Google image.

Below is the identification key for the three species of *Hypophthalmus,* according to Oliveira (1981).

According to Oliveira (1981), the genus *Hypophthalmus has* three distinct forms which, when compared with the literature, have been identified as the species *Hedentatus* Spix, *H. marginatus* Val, and *H. fmbriatus* Kner. These species can be distinguished by the following key:

1 - mouth commissure never reaching the level of the orbit; mentonian barbilla never reaching the end of the pectoral fin, with a relatively narrow fimbria, the width of which rarely exceeds half the diameter of the orbit; caudal fin slightly notched, upper lobe tapered and lower lobe rounded; sometimes truncated; edge of the caudal fin and often of the dark caudal fin *H edentatus* Spix (1829).

1st mouth commissure reaching the middle level of the orbit, inferiorly; caudal deeply furcated 2

2 - mentonian barbels much larger than the others, always extending beyond the end of the pectoralis, with a wide fimbria, whose greatest width is always greater than the diameter of the orbit (usually double), clear caudal and anal borders *H fmbriatus* Kner (1857).

2nd mentonian eyebrow generally smaller than the maxillary eyebrow, with a relatively narrow fimbria, the greatest width of which rarely exceeds half the diameter of the orbit;

caudal and anal with a dark edge *H marginatus* Val. (1840).

1.2 MAPARA BIOLOGY

1.2.1 Food

The maparà, found along the Tocantins River, is concentrated in the lower and middle reaches of the river and refers to only one of the three species of fish belonging to the genus *Hypophthalmus,* the species *Hypophthalmus marginatus* (ALCÂNTARA-NETO, 1994). While most fish are bottom dwellers and carnivores, the maparà feeds strictly on pelagics and plankton (CARVALHO, 1978a). The maparà contributes to the ecological balance by feeding on plankton and serving as food for carnivorous predators. The maparà has practically no competitors for food, unlike almost all commercial species. This gives it a great ecological advantage over other species (ALCÂNTARA NETO, 1994).

The high concentration of maparà in the Cametâ micro-region seems to be related to the numerous river islands that suffer partial or total flooding due to the effect of the daily tides. The constant incorporation of organic material apparently benefits the plankton and detritus that feed the mapara (CARVALHO, 1978a). The three species have a slight difference in diet.

In the Tocantins River, the diet of *H. marginatus* consists of phytoplankton (diatoms and chlorophyceae) and zooplankton (sponge spicules and planktonic crustaceans; copipods, copepod nauplii and rotifers). (CARVALHO *et al,* 1978). The diet of *H. edentatus, as* proposed by Marlier (1967) and Carvalho (1980a), who established that it is a zooplankton species, is: planktonic crustaceans (cladocerca and copepodites and adults and ostracoda); algae, mites and insects (larvae of culicidae, chaoboridae, chironomidae and nymphs of hemiptera) and decapoda (including larval forms of shrimp). The diet of *H. fmbriatus, on the* other hand, is based on zooplankton: cladocerca, copepods and ostracods (CARVALHO & GOUDING, 1985).

1.2.2 Maple's economic importance

In the Cametà micro-region, *H. marginatus* is a fish of great economic value, its catch is considered to be the most significant; it plays an active role in the region's economy, represents one of the main sources of animal protein and is the basis of the diet of the riverside populations. In the municipality of Cametà alone, this species contributes 37% of the annual landing (CARVALHO & MÉRONA; 1986).

The traditional maparâ (*H marginatus)* fishery in the lower Tocantins River basin is considered to be the most important in the micro-region. This fish is part of the essential diet of the riverside population and its economy; it mainly supplies the markets in the municipalities of Abaetetuba, Igarapé-miri, Cametà (where, according to Santos *et al,* 1984, it represents more than 30% of the total catch) and Tucurui (EVANGELISTA *et al,* 1990).

In the Lower Amazon, during the harvest season, fishing for maparâ is an opportunity for employment and a source of income for hundreds of fishermen. However, the maparâ (like other flat fish), perhaps because of the "reima" belief, has a low local consumption; its catch is exported to other Brazilian municipalities and states (ALCÂNTARA-NETO, 1994).

The maparâ has a high meat utilization rate per gross weight of fish. In a 40 cm specimen of *H. edentatus* caught by the Belém fishing fleet, its abdominal cavity does not reach 6 cm in length (OLIVEIRA, 1981). In 1974, the fillet yield was approximately 64% (BRITO *et al,* 1984). The maparà has firm, oily meat, with few fine bones, which are easily removed, with good palatability and aspects (CARVALHO, 1978a). According to Smith (1979), the meat of *Hypophthalmus edentatus* is composed of 28.8% fat, 12.4% protein and 2.6% ash and, according to IDESP (1966), is rich in lipids.

Carvalho (1980b) concluded that females of this species have a higher fat content as the ovaries mature. According to this author, this species does not form a specific fat deposit, but it is diffused in the muscle tissue, which is why the fillet contains a high fat content. The protein content is similar for males and females.

1.2.3 Migration

In the Tocantins River, Carvalho (1978a) observed that maparâs migrate upstream, or reproductively, and downstream, or trophically. These fish are distributed throughout the area, so that from December to February the genetic migration ends and the trophic migration begins when, after spawning, they move downstream in search of food (CARVALHO & MÉRONA, 1986). Carvalho (1978b) assumes that the first spawning occurs when this species reaches a total length of 370 mm and a height of 60 mm.

The mapará depends on the current of the river for fatigue through upstream migration and reproduction. According to Godoy (1979), the excess fats and proteins accumulated need to be metabolized by migration, resulting in muscle fatigue and lactic acid, an important activator of the hypophysis gland and the act of reproduction.

Every year, during the flood season, the fish go up the river and spawn in the headwaters of the small streams of the tributaries of the Tocantins River and in the marginal lagoons located in the municipality of Baiâo, which are flooded at this time due to the surplus of water and the opening of the dam's spillway. When the rainfall decreases, the spillway is progressively closed, making it impossible for water to flow into the stretch of the River Tocantins downstream of the dam. With the rapid lowering of the marginal waters, there is an intense mortality of eggs, larvae and fry, frequently reported by the fishing community. This fact is of the utmost importance when you consider that, every year, juveniles are unable to recruit, which considerably hampers the replenishment of fish stocks and, as a result, there is a stretch of the river that is not very fishable. Only in atypical years, when the dam spillway is open for a longer period of time, do juveniles manage to return to the main bed of the Tocantins River, as happened in 2000, when there was a greater occurrence of juveniles. Efrem & Jansen (2000).

H marginatus lives in the lower Tocantins River, in the 270 km stretch between the hydroelectric dam and its mouth. Its population components remain provisionally separated; the young in the first months of the year are concentrated at the mouth of the Tocantins River or even in a small stretch of the Parà River and the adults mainly in the Baiâo municipality (CARVALHO & MÉRONA, 1986).

From the first stages of life, they group together in homogeneous shoals, with little fluctuation in length. Between March and October, when the Tocantins River is at its lowest ebb and there is probably some influence from the sea at the mouth, there is a slow upward migration in shoals of small (immature) individuals in the main channel of the river at a depth of up to 12 m (CARVALHO & MÉRONA, 1986).

This migratory movement evolved until the vicinity of the subarea of the municipality of Baiâo, known as Mocangui, where the shoals probably disperse in mid-October, when they become difficult to catch (CARVALHO & MÉRONA, 1986).

Shoals of large specimens, belonging to older age groups than those that have dispersed, form in November in the Mocajuba sub-area and above. They make a counter-migratory, reproductive migration and, after spawning, which occurs mainly in January and February in the Baiâo and Tucurui sub-areas, they disperse again throughout the area. This scheme is in line with the observations of Carvalho (1978a), which show the areas and times of occurrence of the shoals.

1.3 FISHERIES LEGISLATION, WITH EMPHASIS ON MAPARA FISHING

The practice of inland fishing is governed by SUDEPE ordinance No. 662, of November 17, 1970, with modifications incorporated by ordinance No. 464, of November 8, 1972, which censors, among other devices, any trawl and cast net (alinea a) and waiting nets with meshes smaller than 70 mm, measured between opposite angles (alinea b). Article 3 defines that during the piracema season, only the use of handlines, single reeds, floats and longlines is permitted.

Article 1 of SUDEPE Ordinance No. N-10, of April 29, 1987, which specifically addresses fishing for the maparâ (Hypophthalmus *spp.) in the* Tocantins River basin, in the state of Parà, prohibits fishing for the species every year from November 1 to January 31. Article 2 effectively bans fishing in breeding and nursery ponds. There is currently Decree No. 132 of October 1, 2001, which sets the closed season for November 1 to February 28. Article 3 defines a minimum total length of 28.4 cm for catching specimens, with the second paragraph allowing 15% of specimens with smaller sizes. Article 4 prohibits the use of trawls or fast cast nets and block nets with an elevation of more than 12m and a mesh size of less than 60mm, measured between opposite angles, throughout the whole of the Pará basin. The sole paragraph of this article states that no fishing vessel may take fishing gear on board that is not in line with the provisions of this article.

The minimum catch size for specimens of *Hypophthalmus spp.* was changed to 29 cm by Article 1 of Ordinance/SUDEPE No. 039 of December 2, 1987, which stipulated the minimum catch sizes for certain specimens in the states of Parà and Amapà. The second paragraph of this article allows for a tolerance of 10% of undersized specimens. In the lower Tocantins, the fishing gear used to catch *H. marginatus* is block nets and trawling poles.

1.4 THE FISHING AGREEMENTS OF THE CAMETA Z-16 COLONY

Fishing and the relationship with the river mark the material and social life of artisanal workers who make sustainable use of its biodiversity as a source of food. Rivers, lakes, streams and boreholes are part of the life of the riverine man of the lower Tocantins and allow him to make a living. The waters are a source of food, products to sell and a means of transportation, keeping the culture of the extractive economy alive. The fishing tackle, such as a malhadeira, matapi, hook and line, and harpoon, which have been used by the river dwellers for centuries, represent this cultural dynamic (PDA, 2006).

This balance was disrupted by the social, environmental and economic impacts of the Tucurui dam, which had a very strong impact on fishing production, threatening the survival of local populations, both urban (consumers) and riverside (artisanal fishermen). In order to remain in their spaces, the communities have developed new strategies for managing their fishing resources, adapting to the new environmental conditions.

One of the initiatives pursued by the fishermen was the conservation of fishing and forest resources through "Fishing Agreements" or "Preservation Agreements". This reflects their concern about the scarcity of fish and the future survival conditions of their families (PDA, 2006).

In the municipality of Cametâ, the idea of agreements began in 1985, with the first experience in the community of Paruru de Janua-Coelis. It was from this initiative that artisanal fishermen began to formulate Fishing Agreements at community meetings, establishing a set of rules with the participation of the entire community and defining rules and defining access to and use of fishing resources.

These agreements, delimited to a river or a place located in a wetland, define rules whose main objective is to reduce the fishing effort by limiting access and how it is used. Limiting access means restricting places where certain types of fishing cannot take place. For example, you can't cast nets in breeding ponds, but you can fish with a reed.

3. MATERIAL AND METHODS

3.1 STUDY AREA

The research was carried out in the city and riverside area of Cametâ, specifically in the community of Jaracuera, an area made up of islands (Figure 4). The municipality of Cametâ is located in the northeast mesoregion of Pará, in the Cametâ micro-region. It has an area of 3022km^2 and is bordered to the north by the municipality of Limoeiro do Ajuru, to the south by Mocajuba, to the east by Igarapé Miri and to the west by Oeiras do Parà (IDESP, 1966). Its geographic coordinates are 2°14'32"S latitude and 49°29'52"W longitude (CARVALHO, 1992).

Three main techniques were used to investigate the current state of maparà *(H. marginatuS)* dynamics: current fish morphometrics, interviews with people involved in fishing activities and a survey of existing literature on the biology and ecology of this species, including studies carried out by Eletronorte in the study area.

Figure 4: Location of the Jaracuera community.

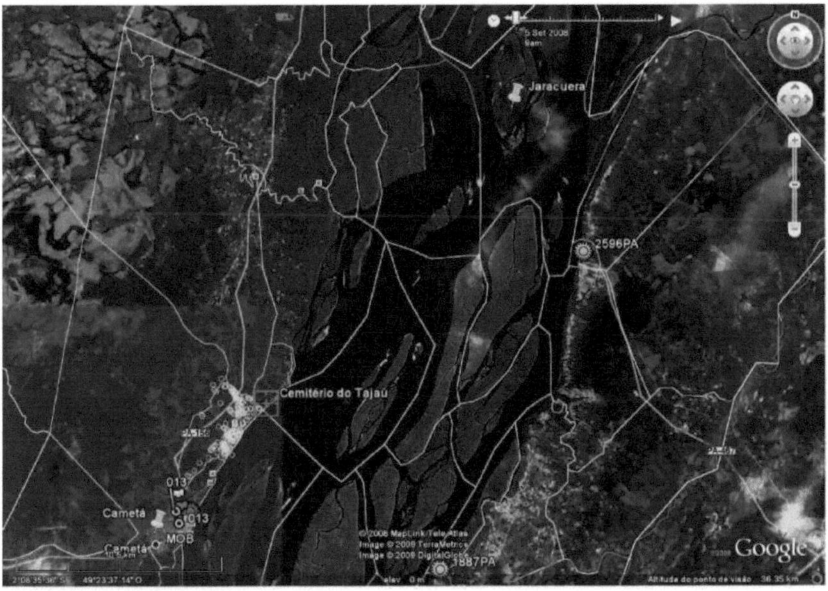

3.2 BIBLIOGRAPHIC DATA collection

The research consisted primarily of scientific data collection, where the various pieces of information were obtained through different types of literature, such as books, articles, etc. In addition to these, searches on Internet sites were used as a complement. Some data was obtained directly from: IBAMA, IBGE, the Federal Rural University of Amazonia - UFRA, PDA, MPEG and ELETRONORTE.

3.3 FIELD DATA COLLECTION

The field research was carried out in two stages: The first consisted of informal and formal interviews with fishermen and fish sellers. The second stage was the collection of morphometric data from specimens of mapará *(H. marginatus)* acquired at Cametâ's fish markets. These measurements were necessary in order to confirm the "information" that the mapará was being caught below the size allowed by law. The interviews were conducted "face-to-face", i.e. verbally with the fishermen, where the questions were prepared in advance, and at times it was necessary to use a small tape recorder to record some information. During this time, we carefully tried to avoid making the interviewees feel uncomfortable, so that their answers wouldn't be false or omitted, since they are extremely suspicious.

During the visit to the Jaracuera community, five fishermen were interviewed, plus the president and the coordinator of the community's "fishing agreement". In the city of Cametâ, the president José Fernando and agronomist Waldir of the Cametâ Fishermen's Colony Z-16, the IBAMA coordinator and an Eletronorte employee responsible for collecting data on fish landings in the city of Cametâ were also interviewed.

3.3.1 Morphometry

To obtain the morphometric data, 103 specimens of maparâ *(H. marginatus)* were randomly measured at the fish market and free market in the municipality of Cametà during six consecutive working days in October 2008. The measurements were carefully taken in the afternoon, considering that during this period fishermen sell juvenile fish called "miùdos", which is forbidden by law. The specimens were measured using a handmade ichthyometer, where the CF, the length from the tip of the snout to the notch in the caudal fin, was obtained. All measurements were recorded in a field data collection table and statistically analyzed (Figure 5).

Figure 5: Morphometry of a specimen of *H. marginatus,* measuring approximately 15 cm.

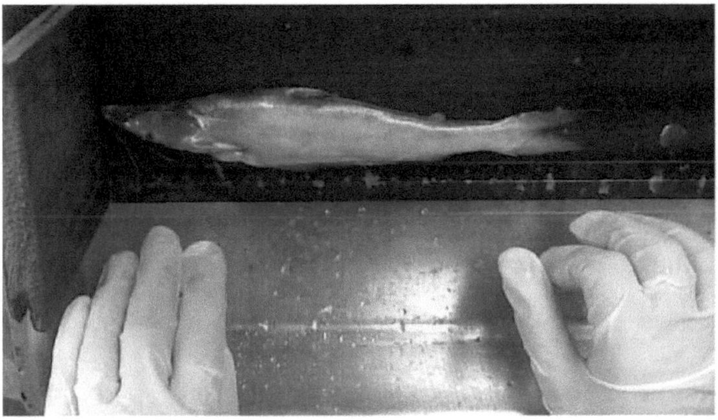

Source: Viana, L. L.

4. DATA ANALYSIS

For the analysis of the interviews, we tried to observe the coherence of the fishermen's answers and compare them with existing scientific information.

The number of individuals and the respective morphometry records were grouped into length classes 12 I- 14; 14I - 16; 16 I- 18; 18 I- 20; 20 I- 22; 22 I-I 24, stored in an Excel 2003 spreadsheet, from which quantitative relative frequency graphs (%) were generated for the groups.

5. Results and discussion

Of the three known species of Mapará, only *Hypophthalmus marginatus* occurs in the Tocantins River, but its external morphological characteristics showed some differences in shape and coloration, according to the habitat in which they are found. These different characteristics have led to the hypothesis that they may be two distinct species of mapara and not just one. This subsidy leaves an open door for researchers interested in continuing their research.

The mapará (*H marginatus)* found along the route from the mouth of the Parà River to near the Tucurui dam is popularly known as the "mapará-do-nosso" (Figure 6), and is preferred by the Cametaenses and also by the municipalities located along the river. They are caught in lentic environments, i.e. those with a current. They are smaller, robust and have a distinctive coloration, with a lighter loin. For the local population, this fish has a more pleasant taste and is less "pitiù" than the other fish found both along the course of the Tocantins River above the dam and in the dam lake, i.e. in lotic environments, still water. The mapará caught in still waters has peculiar characteristics, its body is more elongated and narrow and it has dark loins, being popularly known as "Paulo Isidório", (in homage to the black-skinned former player of the Brazilian national team) (Figure 7), these, however, are more pitiù and less pleasant in taste than the "mapará-do-nosso".

Figura 6: Species *H. marginatus,* known as the "mapará of ours"

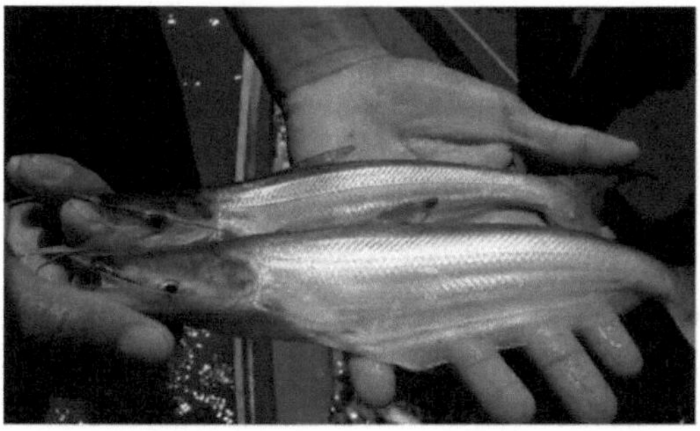

Figura 7: Species *H marginatus,* known as "maparà Paulo Isidório"

Source: Viana, L. L.

During the research it was observed that the popular preference of Cametaenses is for the so-called "maparâ-do-nosso", scientifically known as *H. marginatus,* which leads many fishermen to commit infractions by fishing during the species' closed season. At the fish market in the municipality and at some of the open-air fairs, a significant number of fish were sold, ranging in size from 12 to 24 cm (Table 1), popularly known as "fifites" (Figure 8), well below the 37 cm length required for the fish to mature and reproduce, according to the size found by Carvalho (1978b). The Cameta population's habit of eating smaller specimens is due to the fact that they believe that adult maparâs are "rowdy" and juveniles are not. This is known as "popular belief". It is also worth noting that the "fifite" fish found were all the so-called "maparâ- do-nosso" and three times the size of the "maparà Paulo isidório" at over 37 cm in length. This is another crucial factor in the significant decrease in the maparà ("maparâ-do-nosso") population, as they are unable to recruit.

Another important factor is the increase in the population of the riverside area. In the 2000 IBGE census, the population of Cameta was 97,504. The population living in the urban area corresponds to 41.42% and the rural area to 58.58%. More than 50% of the rural population lives in the island region, which is made up of 122 localities. There are approximately 35,000 people, known as ribeirinhos (river dwellers), who look to the river for their livelihood. In the last census in 2007, the population was not broken down by zone, but a total population of 110,323 was recorded. And according to the Z-16 Fishermen's Colony, Cametà has a total of approximately 15,000 fishermen, with only 7,500 associated with fishing agreements. This increase in the riverside population increases the predatory fishing of maparâ, both adults and juveniles, because as the big fish dwindle, fishing turns to the "smaller" fish.

Figura 8: A specimen of *H. marginatus,* known as "fifiti".

Source: Viana, L.L

As the data in Table 1 shows, artisanal fishing for the maparâ *(H marginatus)* in the Municipality of Cametà poses a significant risk to the species in question, where the stock could enter a state of overfishing and very probably cause serious bio-ecological changes in the area in which it operates. This significant proportion of juvenile individuals caught throughout the year could reduce the maparâ stock during fishing trips, causing serious damage not only at a bio-ecological level, but also at a socio-economic level, since the Cametaense population depends fundamentally on fishing in the Tocantins River to continue supplying the domestic fish market and offering employment to a significant contingent of labor, especially with maparâ fishing, which is traditionally the popular preference.

Table 1: Shows the number of individuals of the species *H. marginatus,* by size category in length classes (cm).

Length class (cm)	Number of individuals	Min.	Doctor.	Max.	Grand total (%)
12l-14	6	12	12,75	13,5	5,8
14l-16	14	14	14,75	15,5	13,59
16l-18	23	16	16,75	17,5	22,33
18l-20	20	18	18,75	19,5	19,41
20l-22	21	20	20,75	21,5	20,38
22l-l24	19	22	22,75	24	18,44
TOTAL	103	-	-	-	100%

The length frequency of the 103 individuals measured of the *H. marginatus* species ranged from 12 to 24 cm in total length, as shown in Graph 1. The highest frequency of catches during fishing trips is with individuals measuring between 16 and 18 cm in length (Graph 1). Based on the data from this survey, it was also observed that the average number of individuals in all size classes was 18.53. The number of individuals per size class was also very representative in the 16 to 18 cm and 20 to 22 cm class intervals, see Graph 1.

Graph 1: Shows the number of individuals by size class

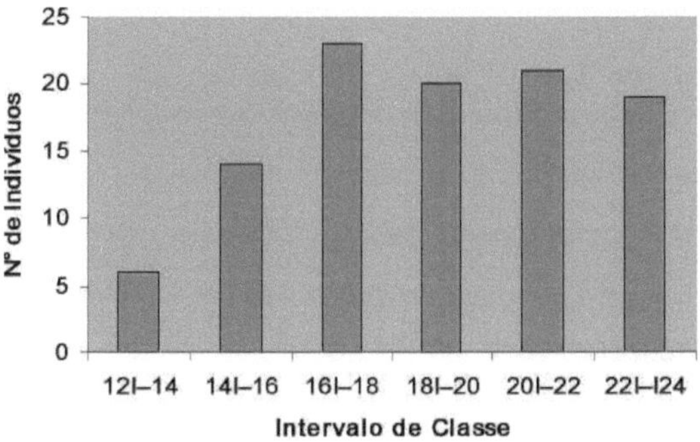

Graph 2: Shows the representation (%) of individuals by size class

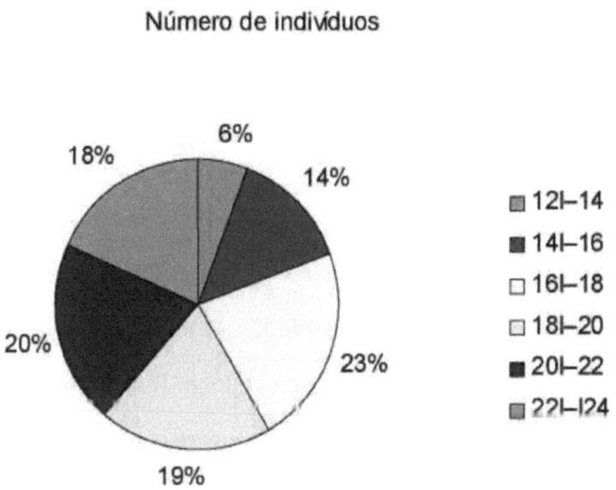

Based on scientific data and information acquired from experienced fishermen in the rogion, it was also observed that the Tucurui HPP has a significant influence on the behavior of the mapara *(H. marginatisi). This is* due to the fact that this species spawns in deep areas of the river that are more than nine fathoms deep, which are known as pools or cucuias. According to Mérona (1986), who carried out research before the completion of the Tucurui HPP, he stated that the dam would not affect the population of the mapara *(H marginatisi* since the fish does all its genetic migration in the Mocangui region, located just above the town of Baiâo and below the dam (Figure 9). In practice, however, this has not been the case, because with the damming of the Tocantins River, it has been observed that the water level of the river downstream has been greatly reduced, thus affecting the supposed reproduction and spawning area of the mapara fish *(H marginatisi), as* well as its feeding habits, since the fish sometimes needs to seek food on the banks of the river, through the riparian vegetation.

> "The problem with the dam was that when they migrate, it's not in the dry season, they used to go deep, they used to go to the puçu, now there's no puçu that deep anymore, because when the dam closed it dried up for us, and the river dried up all over, so there's nowhere for them to reproduce anymore, and that's why they've decreased." (J.M. M. F.)

Although Mérona (1986) mentioned in his work that maparà spawning occurs throughout the Mocangui region, reports from fishermen with thirty years' experience say that spawning occurs throughout the region downstream of the river, due to the high concentration of fry and ovulated females during fishing trips in the region's cucuias (Figure 10).

> "You know why, when the fishing season closes, for those of us who have experience, this material here is a predator in the lowlands, when it goes down (the mapara) it goes down ovate out there, so it's not just up there that it spawns, we catch it ovate big." (J. M. M. F)

> "There are some big cucuias here, examples: in Capim, Ilha de Baiâo, Cipoteua and Praia Grande, they are big cucuias, really tough, very good." (J. P.)

Figura 9: Spawning site of the mapará according to Carvalho & Merona (1986)

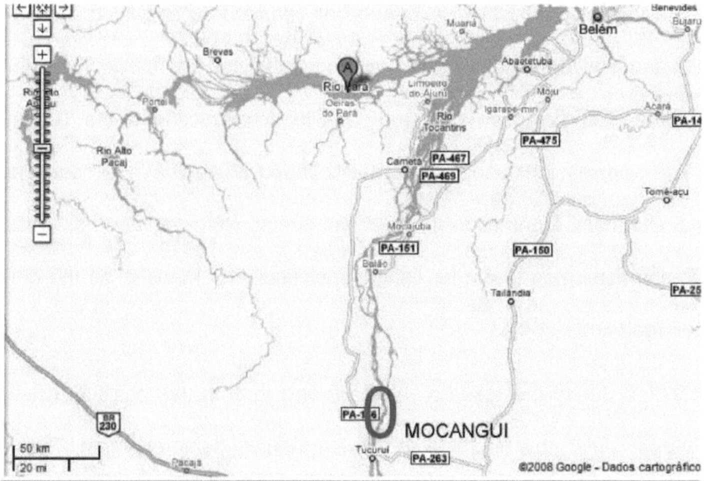

Source: Google - cartographic data, 2008

Figura 10: Map showing the location of the maparà spawning area, following the Fishermen

Source: Google Earth, 2008

On the basis of these two pieces of scientific information and "face-to-face", we were led to the conclusion that the maparà must have expanded or changed its spawning grounds after the changes in the environment, because in Mocangui, the fish stock has been greatly reduced. According to the local fishermen, with the construction of the Tucurui HPP, most of the fish species that used to be easily found along the river, such as Piranha, Branquinha, Curimata, Mantrichao, Jaraqui and Aracù, were retained above the dam. This has led many fishermen to opt for fishing specifically for maparà, as the only source of fish for their food and income.

The interviews conducted during the research also showed that, in the localities where there are "Fishing Agreements", this shortage of fish is gradually being changed. This is due to the fact that the fishermen have changed their working dynamics, such as banning the fishing of small fish and the use of trawlers, the use of block nets with a mesh size of more than 6 cm and up to nine fathoms deep, and the closure of fishing during the closed season, which runs from November 1st to February 28th or 29th. To prevent predatory fishing during the closed season, in the area where the "agreements" are in force, the fishermen are divided into groups to monitor the river at night; when someone is caught fishing, the fisherman's work equipment is seized. All these measures are making the maparà *(H. marginatuS)* reappear in these preservation areas, but unfortunately these measures are only practiced in these localities, and maparà fishing during the closed season and the "Fifite" along the Tocantins River is still very high.

6. FINAL CONSIDERATIONS AND RECOMMENDATIONS

The mapará population in the lower Tocantins River is increasingly at risk, mainly due to the fishing lifestyle of the riverbank dwellers; they are fishing, even with the closed season, during the spawning season, in addition to the large quantities of "fifites" fish that are caught, due to the widespread use of predatory fishing tools. The fact that this situation continues, even after the creation of the laws, is due to the fact that the laws are not being rigorously applied by the competent bodies. However, artisanal fishing in Cametâ needs to be adjusted to the current pattern of behavior of the mapará stock, so that exploitation can find the desired level of balance and its continuity can be guaranteed;

Bearing in mind that artisanal fishing for mapará is a means of subsistence for many families, especially riverside communities, because it supplies the domestic fish market in certain regions and, above all, because it has a socio-economic and cultural value for the municipality, it is urgently important to extend the **"Fishing Agreements" to** all riverside communities, since they are producing results in the communities that practice them. Consumers should also be more aware of the possible aggravating factors that could arise when buying from the "fifite" market.

The elaborate interviews led to reports of the capture of 25-cm-long ovate mapará, which is below the 37-cm size stipulated by Carvalho (1978b) for first reproduction, and also because we find "fifites" at any time of the year, especially in November, which led us to two questions:

1. - Due to the low availability of food and the reduction in the volume of the river, influenced by the insertion of the Tucurui HPP and/or even the destruction of the riparian forest, some specimens of the mapara are not reaching the size proposed by Carvalho (1978b), but this does not affect their reproduction.

2. - The large size varieties of Mapará sold close to the period of genetic migration

raises questions about the reason for this variety, whether Maparà spawn throughout the year or whether they need more than one year to reach breeding age;

However, precise scientific information on their bio-ecological behavior must be established as a matter of urgency, paralleling the results of previous research to see if there are any changes.

We suggest:

3. Formulation/execution of a maparà research program to cover its entire distribution area, particularly along the Tocantins River, and to answer all outstanding questions, such as breeding times and places, average size at first sexual maturation, reproductive potential, bio-ecological behavior, recruitment pattern, stock potential, etc;

4. Immediate resumption of statistical control of artisanal maparà production, especially in the Cametà region, where consumption of this fish is significant;

5. That the competent bodies, such as IBAMA, for example, increase the inspection service and that the punishments are severe for those who commit the infractions, as stated in the laws in force;

6. That Cametà's own municipal bodies, such as SEMMA, mobilize and form a partnership with the Fishermen's Colony Z-16, expanding inspection and awareness-raising services.

The lack of scientific knowledge on the part of the local population leads them to commit offenses. It is suggested that the results of the monthly catches made by the competent bodies be passed on to them, showing what the undue exploitation of the maple stock could cause to the environment, society and the local economy. To this end, it is recommended that lectures and booklets be given in accessible languages, with the aim of raising awareness among those who fish; those who sell; those who consume; and those who fish/trade/consume at the same time.

It is worth noting that the maparà fishing area in the lower Tocantins lacks studies on the anthropological, cultural, social, economic and ecological aspects of the artisanal fishermen.

7. REFERENCES

AGOSTINHO, A. A. **Management of fish resources in reservoirs**, p. 106-121, In. AGOSTINHO, A. A; BENEDITO CECILIO, E., Current situation and perspectives of Ichthyology in Brazil. Doc. do IX Encontro Brasileiro de Ictiologia. Maringà: ed. UEM, Maringà, eds. (1992).

ALCÂNTARA NETO, C. P. **Ecologia da pesca dos Maparas, Hypophthalmus spp. (siluriformes, Hypophtalmidae), no lago Grande de Monte Alegre, Baixo Amazonas, Para**. Master's dissertation. Federal University of Pará/ Emilio Goeldi Museum of Pará. Belém, Pará. 144 p. (1994).

BOHLKE, J. E.; WEITEKAN, S. H.; MENEZES, N. A. **Current status of freshwater fish systematics in South America**. *Acta Amazon.* 8(4): 657-677, 1978.

BRITO, R. C.C. et al. **Entrepreneurial fishing in Para**. Belém: IDESP, 1975. 72 p.

BRITISK, H. A.; SATO, Y. & ROSA, A. B. S. **Manual de identificaçâo de peixes da regiâo de Très Marias.** Brasilia: Câmara dos Deputados/CODEVASF, 143 p.

(1984).

CARVALHO, D. M. **Artisanal fishing in the municipality of Cameta.** UNESPA: Belém. Bachelor's thesis. 82 pages. 1992.

CARVALHO. F. M. **Composiçâo quimica e reproduçâo do mapara *(Hypophthalmus edentatus* Spix, 1829) do lago Castanho, Amazonas (Siluriforme, hypophthalmidae).** Acta anazon. Manaus, V.20, n.2, p. 379-389 (1980b).

CARVALHO, J. L. **Contribuiçâo ao conhecimento da biologia do mapara *(Hypophthalmusperporosum).* COPE 1878 (pisces: Hypophtalmidae) in the lower and middle Tocantins.** - Bol. FCAP 10: p.39-57. (1978a).

CARVALHO, J. L. **Selectivity of the main devices used to catch mapara**

***(Hypophthalmusperporosum)* COPE 1879 (pisces: Hypophtalmidae) in the lower and middle Tocantins**. - Bol. FCAP 10: p.77-92 (1978b).

CARVALHO, J. L. & MÉRONA, B. - **Studies on two migratory fish from the lower Tocantins before the closure of the Tucurui dam, *Amazoniana*** 9 (4): p. 595-607. (1986).

CARVALHO, M. L. & GOULDING, M. **On the feeding ecology of the catflish *Hypophthalmus fimbriatus* in the blackwater Rio Negro of the Amazon Brasin.** Rev. Bras. Zool. Sâo Paulo. V. 3n.1, pg 33-41. (1985).

CET - Consórcio Engevix-themag. **Interpretative report on limnological characteristics (May/85-April/88).** Brasilia: Consórcio Engevix-themag, 1988.

CLARK, C. W. **Mathematical bioeconomics: the optimal Management of renewable resources**. 2nd Ed. John Wiley & Sons. Xiii+386p. (1990).

COLLART, O. D. **Ecological studies and environmental control in the region of the Tucurui HPP reservoir; shrimp production segment in the lower Tocantins.** Sector report, ELN/CNPq/INPA, 1986.

CUMMINS, K. W.; MINSHALL, G. W.; SEDELL, J. R., CUSHING, C. E & PETERSON, R. C. **stream ecosystem theory**. Verh. Internat. Limnol. 1984. 22: p. 1818-1827.

CURY, P. **Une approche théorique de l' impact de l' environnement sur la pêche**. In: CURY, P. & ROY, C. *o.uesr - Africaines pêcheries. Variability. Instabilité etChangunent* ORSTOM, p. 368-376, eds. (1991).

EFREM, J. G. F. & JANSEN, A. S.Z. **Ichthyofauna of the Tucurui HPP, Tocantins River, World dam commission: Brazilian case studies, Tucurui HPP, final report**. - annex 6, p. 106-121. (2000).

EVANGELISTA, J. E. V. et al. **Relatório de estudos sobre a Pesca do Maparà**, Belém,

IBAMA/SEPSUL/SUPES-PA. 14p. 1990. mimeographed.

FEARNSIDE, P. M. **Social impacts of Brazil's Tucurui Dam. Enveronmental Manogement**, 24 (4): pag 483-495. (1999).

FERREIRA, E.J.G.; ZUANON, J. A. S. & SANTOS, G. M. **Peixes comerciais do mèdio Amazonas: regiao de Santarém, Parà.** Brasilia: IBAMA, 214p. (1998).

GODOY, M. P. **Marking and migration of the piramutaba *Brachyplatystoma vaillant!!!* (VAL. 1840) in the Amazon basin (Parà and Amazonas), Brazil (pisces, mematognathi, Pimelodidae).** - Bol. FCAP 11: 3-21. (1979).

HOWELL, W. M. & BLACK, D. A. **Controlled silver-staining of nucleolus organizer regions with a protective colloidal developer: a 1-step method.** Experientia, 36: 1014 - 1015. (1980).

IBGE. **Fishing statistics, 1989. Major regions - federation units**. Rio de Janeiro, v.10, n.2, july - december, 1989, p.62.

IDESP. Fishing in Para: An open and investigative sector. In. **MEETING TO ENCOURAGE THE DEVELOPMENT OF THE AMAZON**, 1.1966. Manaus. Annals: Belém, IDESP, 1966. p 1-26.

NATIONAL INSTITUTE FOR RESEARCH IN THE AMAZON (INPA). **Ecology and environmental control studies in the Tucurui** HPP **reservoir region**: ichthyofauna, ELETYRONORTE/CNPq/INPA agreement, Brasilia. Unpaginated (report), 1986.

JURAS, A. A. CINTRA, I. H. A.; LUDOVINO, R.M.R. A **pesca na área de influência da usina hidrelétrica de Tucurui, Estado do Parà**. Boi. Téc. Cient. CEPNOR, Belém, V, 4, n.1, P. 77-88. (2004).

LANG, C. **Fishing in the Amazon in danger: The situation of the Maparà in Cametà.** Cornell University: Florida- USA. 1998. Translation: Pompeu, J. D. P.; 25 pag. (1998).

LUNDBERG, J. G. & LITTMANN, M. W. **Family Pimelodidae (Long- whiskered cat fishes)** In: REIS. R.E.; KULLANDER, S. O. & FERRARIS, Jr., C. J. eds. **Check list of the freshwater fishes of south and Central America**. Edipuers, Porto Alegre. P. 432-445. (2003).

MÉRONA, B. **Les peuplements de poissous et la pêche daus lê Bas Tocantins avant la fermeture du barrage de tucurui**. Verh. Internat. Verein. Lonenol, 22: 2698-2703, 1985.

MÉRONA, B. **Ecological aspects of the ichthyofauna in the Lower Tocantins.** *Acta Amazon.* **(16/17): 109-124, 1986**

MÉRONA, B. **Amazon fisheries: general characteristics based on two case studies.** Interciencia. 15 (6). Pag 461-468, 1990.

MÉRONA, B. **Ecology of fishing and fisheries management in the Amazon region**. - Bol. Mus. For. Emilio Goeldi, ser. Antropl. 11(2): P.167-182. (1995).

MYERS, G. S. **Usage of anadromous, catadromous and allied terms for migratory fishes**. - Copeia: P. 89-97. (1949).

NELSON, J. S. **Fishes of the world.** 3. ed. New York: John Wiley & sons, (1994).

OLIVEIRA, J. C. **Osteology and systematic review of the Hypophtalmidae family (Teleostei, Siluriformes).** Sâo Paulo: 101p. Dissertation (Master's degree in science - zoology). University of São Paulo/Institute of Biosciences (1981).

PEDREIRA, J. M. M. **Dicionârio de Peixes de Couro do Brasil**. Belém: SUDAM, 122P. P.60. 1971.

PETRERE JR. **Fishing in the Amazon.** In: SEMINARIO INTERNACIONALSOBRE AMBIENTE, POBREZA E DESENVOLVIMENTO DE AMAZONIA. Belém. Proceedings. Belém: Governo do Estado do Para, Secretaria de Estado de Ciência, Tecnologia e Meio

Ambiente, p. 72-78. (1992).

PDA - Cametâ. Cametà: **Fishing agreements - an economic and organizational alternative**. Cametà. Anâs. Cametâ: Federal Government, Ministry of the Environment, 2006, 64p.

RAPP PY-DANIEL, H. L. & FERNANDES, C. C. **Sexual dimorphism in Silurformes and Gymnotiformes (Ostariophysi) from Amazonia.** Acta Amazon. Vol. 35(1), pag. 97-110 (2005).

REIS. R. E.; KULLANDER, S. O. & FERRARIS, Jr., C. J. **Check list of the freshwater fishes of south and Central America**. Edipuers, Porto Alegre. 729p. eds, 2003.

ROBERTS, T. R.: **Ecology of fishes in the Amazon and Congo Basins**. - Bull. Mus. Comp. Zool. 143 (2): 117-147. (1972).

SANTOS, G. JEGU, M. MÉRONA, B. **Catalogo de peixes comerciais do baixo Rio Tocantins**. Manaus: ELETRONORTE/CNPq/INPA (1984).

SANTOS, G. JEGU, M. MÉRONA, B. **Catalogue of commercial fish from the lower Tocantins**. Pub. ELETRONORTE: 83p. (1985).

SANTOS, G. M. **Peixes comerciais de Manaus / Geraldo** dos Santos, Efram, J. G. F, Jansen, A. S. Z. - Manaus: IBAMA/AM, Provàrzea. 144p. (2006).

SILVA, R. C. **Mapara Fishing**. Cordel Literature. Belém. Anas. Belém: government of the state of Parà, secretariat of culture, sport and tourism, 1986. P. 1011.

SMITH, N. J. H. **Fishing in the Amazon River**. Manaus, CNPq/INPA, 154p. (1979).

WOOTTON, R. J. **Introduction: Strategies and tactics in fish reproduction**. In. Potts, G. W.; WOOTTON, R. J. **Fish reproduction: Strategies and tactics**. LONDON: ACADEMIC Press. P. 1-12. 1984.

WORD, J. V. & STANFORD, J. A. **Riverine ecosystems: the influence of man on**

catchment dynamics and fish ecology. In: Dodge, D.P. (ed.) *Can. Spec, publ Fish. Aquat. Sep* 106: 56-54. (1989)

MIX
Papier aus verantwortungsvollen Quellen
Paper from responsible sources
FSC® C105338

Printed by Books on Demand GmbH, Norderstedt / Germany